Isaac Newton

Illuminating the Wonders of Motion and More

Michael Stamp & Renée Dusseau

ICR
INSTITUTE FOR CREATION RESEARCH
Creation Kids Books

Dallas, Texas
ICR.org

Isaac Newton
Illuminating the Wonders of Motion and More

by Michael Stamp & Renée Dusseau

First printing: March 2024

Designed by Susan Windsor
Contributions by the ICR Communications team: Jayme Durant, Beth Mull, Lori Fausak, Bethany Trimble, Rachel Brown, and Frank Sherwin

ISBN: 978-1-957850-42-9
Library of Congress Control Number: 2024934622

Please visit our website for other books and resources: ICR.org

Printed in the United States of America.

Published by Creation Kids Books, an imprint of ICR Publishing Group

ICR PUBLISHING GROUP

Table of Contents

"This most beautiful system of the sun, planets, and comets could only proceed from the counsel and dominion of an intelligent and powerful Being."
— Isaac Newton

Foreword

Isaac Newton is featured in ICR's Heroes of Creation Science series because he was a scientist who studied the Bible and trusted in God as the Creator of the universe. He remained dedicated to exploring the great mysteries of God's creation throughout his life.

Gravity is one of many discoveries that made Newton famous. This force was designed by the Lord Jesus Christ to hold everything in our world together. In fact, Hebrews 1:3 says He upholds "all things by the word of His power." God is mighty to sustain the entire cosmos—and that includes you, too! He keeps every detail in place out of His great love for you.

As you read about Newton's incredible theories of gravity, light, and more, you can remember that among all the wonderful things God made, you're His most special creation.

Isaac Newton 1643–1727

Introduction

Isaac Newton is known as one of the most brilliant scientists in history. From an early age, he demonstrated the curiosity and intelligence that God had given him. Though Newton was a very private man and worked alone for most of his life, he felt a strong duty to share his pursuit of truth, discoveries, and research with others.

Newton was part of the Scientific Revolution, a period during the sixteenth and seventeenth centuries when people changed how they viewed the world. This time was filled with all kinds of new ideas about science and the way it's studied.

Newton built his research on the foundation of many great scientists who came before him. He described his three laws of gravity and motion in his 1687 book *Principia Mathematica*, which provided the basis of modern physics. But Newton wasn't just a physicist. He also studied math, astronomy, chemistry, optics, and theology, among other subjects.

Above all, Newton dedicated his studies to God. He hoped they would help people learn more about Him. Through exploring the mysteries of God's universe, Newton believed that science points us to the majesty of our glorious Creator.

Did You Know?

During Newton's time, the Julian calendar was being used in England. According to that, Newton was actually born on December 25, 1642—Christmas Day!

Newton's birthplace—Woolsthorpe Manor in Lincolnshire, England

Chapter 1

Budding Scholar

Isaac Newton was born in the small village of Woolsthorpe, England, on January 4, 1643. His mother, Hannah Ayscough, had him three months early, so he wasn't expected to live. The sickly baby was tiny enough to fit inside a quart mug.

Isaac never met his father, a farmer

who died before he was born. When Isaac was three, his mother remarried a wealthy minister named Barnabas Smith and moved away. His stepfather didn't want a stepson, so Isaac was raised by his grandmother. His mother had three children in her second marriage, but Isaac didn't grow up with his half-siblings.

Isaac Newton's sundial on display in St. John the Baptist Church in Colsterworth, England

Despite his lonely childhood, Isaac was fascinated with nature and how things worked. He asked a lot of questions about the world around him. The young Isaac also loved to make things, like a sundial that he carved with a pocketknife. One day, after he saw a windmill being constructed, he even rushed home to build a small working version of his own.

A windmill in Holland

Grammar School

At 12 years old, Isaac became a student at Free Grammar School in Grantham, England. He studied classical subjects like Latin and Greek. He also lived with a local pharmacist and studied his chemistry books.

By this time, Isaac's mother had returned following the death of Reverend Smith. She wanted Isaac to take care of the family's property, so she withdrew him from school. But the teenager wasn't interested in agriculture—he preferred to draw, read, and build things.

It soon became clear that Isaac wasn't meant to be a farmer, so his mother allowed him to finish his education. His passion for

Free Grammar School, today called The King's School

Did You Know?

Isaac wasn't a very good student at first. After a fight with one of his classmates, he became determined to try harder in school. By the time he graduated, Isaac had climbed to the top of his class!

learning would motivate his studies for the rest of his life.

Trinity College

In 1661, Isaac enrolled at Trinity College, Cambridge, one of the oldest universities in the world today. It was expensive to go to school, so to help pay his tuition, Isaac worked as a subsizar—a servant like a butler. Before long, he received a scholarship that allowed him to earn his bachelor's degree.

Trinity College, Cambridge, was founded by King Henry VIII.

At the time, most schools in Europe still taught old ideas and beliefs from ancient Greece. The newest discoveries in science and mathematics weren't part of Isaac's education. So, the young scholar decided to investigate these concepts on his own. He recorded his thoughts in a notebook titled *Certain Philosophical Questions*, which laid the foundation for his future work.

> "If I have seen further it is by standing on the shoulders of giants."
> — Isaac Newton

The Shoulders of Giants

Since the discoveries of many influential scientists weren't taught at the University of Cambridge, Newton's reading outside of school would become very important to his future.

Nicolaus Copernicus

Nicolaus Copernicus, for example, championed the idea that planets in the solar system orbit the sun. Johannes Kepler described how they move, and Galileo Galilei used telescopes to see objects in space better.

René Descartes

Newton also read the writings of French philosopher and mathematician René Descartes. His ideas about geometry and algebra helped Newton discover calculus. These scholars were the "giants" whose findings inspired Newton's research.

The Great Plague of 1665-66

From 1665–66, the bubonic plague spread across England. It was the worst outbreak of plague since the Black Death in 1348. Most wealthy people fled the city for safety in the countryside. The nation's borders were closed, and many workers lost their jobs.

The plague spread through fleas that were carried by rats. Since the impoverished areas of London had many rats and people lived close together, the poor suffered the most. After two years, about 15% of the city's population had died from the disease.

The Christians Refuge—*this strange book claimed to contain remedies for the plague.*

Chapter 2

Questions to Answer

In April 1665, Newton graduated from Cambridge with his bachelor's degree. Shortly afterward, the school closed its doors because of the bubonic plague. Newton returned to the safety of home, where he studied on his own for about two years.

During this period, Newton developed several theories in science and mathematics, but he didn't tell anyone what he was doing. Known as his "Year of Wonders," Newton's time in isolation led to many of his important findings.

The Famous Falling Apple

There's an old legend that an apple inspired one of Newton's most valuable discoveries—gravity! As the story goes, the young physicist was sitting under an apple tree when an apple bonked him on the head. While Newton wasn't actually hit on the head, the tale of the falling apple is probably true.

Many years afterward, Newton told this account to the historian William Stukeley, who later wrote a biography of

An apple tree on the grounds of Woolsthorpe Manor, Newton's childhood home

The International Space Station

Did You Know?

Many people believe the famous apple tree is still at Newton's childhood home today. In 2010, NASA astronaut Piers Sellers took a piece of this tree into space. The wooden sample floated inside the International Space Station.

the scientist's life. Newton explained that the apple led him to wonder why objects fall straight to the ground—and he was determined to find the answer.

Science and Faith

From early in his scientific career, Newton recognized that the universe was created by God. He wrote, "All that diversity of natural things which we find, suited to different times and places, could arise from nothing but the ideas and will of a Being necessarily existing." In fact, Newton firmly believed that to reject Him was crazy!

Newton never stopped reading his Bible, and his pursuit of truth motivated his life's

work. He once said, "When I wrote my treatise about our system, I had an eye upon such principles as might work with considering men for the belief of a Deity, and nothing can rejoice me more than to find it useful for that purpose." In other words, Newton hoped his research would lead others to the knowledge of their Creator.

Statue of Isaac Newton studying an apple at the Oxford University Museum of Natural History

The Pull of Gravity

Gravity is an invisible force that pulls things together, even when they're not touching. It keeps us grounded on Earth and holds our entire solar system in place. The planets, moons, and other celestial bodies are locked in perfectly balanced orbits around the sun because of its huge gravitational pull.

On Earth, gravity also explains the predictability of ocean tides. The gravity of the sun and moon pulls on Earth's seas, causing the tides to rise and fall each day. Newton's discoveries influenced how scientists would understand gravity for years to come.

Chapter 3

A Young Genius

The horrible plague eventually ceased, and the University of Cambridge reopened in 1667. Newton returned to college as a professor and earned his Master of Arts degree. Then, he was named Lucasian Professor of Mathematics in 1669. He was the second person to hold this title, one of the most esteemed positions in the world.

The previous chairholder, Isaac Barrow, was a professor of Newton's at Cambridge. He served as a mentor to the young scholar and encouraged his academic pursuits. Barrow eventually left his teaching job for a ministry career, but not before recommending Newton as his successor.

When Newton came back to school, he brought with him calculus, a subject of math he had discovered and worked on during his months at home. Newton also introduced his

Did You Know?

Although Newton was a well-known teacher, not a lot of people attended his lectures. His classes were likely difficult for students to understand. But Newton didn't mind all that much—he preferred to focus on his research, anyway.

theories of light and the motion of planets. It didn't take long for his coworkers to recognize him as a genius.

Calculus Conflict

During the plague epidemic, Newton became one of the first people to study calculus. This is a branch of math that involves complex calculations like areas of curved shapes. It's often used to determine rates—the speed at which things happen—and to describe how objects move. Although Newton wrote several papers on his findings, he didn't publish them until years later.

Around 1674, the German mathematician Gottfried Leibniz also developed a form of calculus. Strangely enough, these men pioneered

Gottfried Wilhelm Leibniz

calculus around the same time. Leibniz and Newton both claimed to have discovered this new math, and the conflict between their supporters lasted for decades.

Reflecting Telescope

In 1668, Newton created a new kind of telescope called the Newtonian reflector. About 60 years before, Galileo Galilei had pioneered the use of telescopes in astronomy. He built a refracting telescope based on descriptions from Dutch spectacle-maker Johann Lippershey. It used lenses that produced a blurry view, but it was good enough for Galileo to see Jupiter's four largest moons and craters on Earth's moon.

Illustration of Newton's reflecting telescope

Newton's reflecting

telescope brought things into sharper focus. Reflecting telescopes are often larger than refracting telescopes, so they collect more light. When light enters the telescope, it first bounces off a large, curved mirror. Then, it hits a small, flat mirror that reflects the rays to the eyepiece. Because reflecting telescopes use mirrors instead of lenses, they provide a clearer view of faraway objects in space. Newton's revolutionary concept became a model for the telescopes scientists use today!

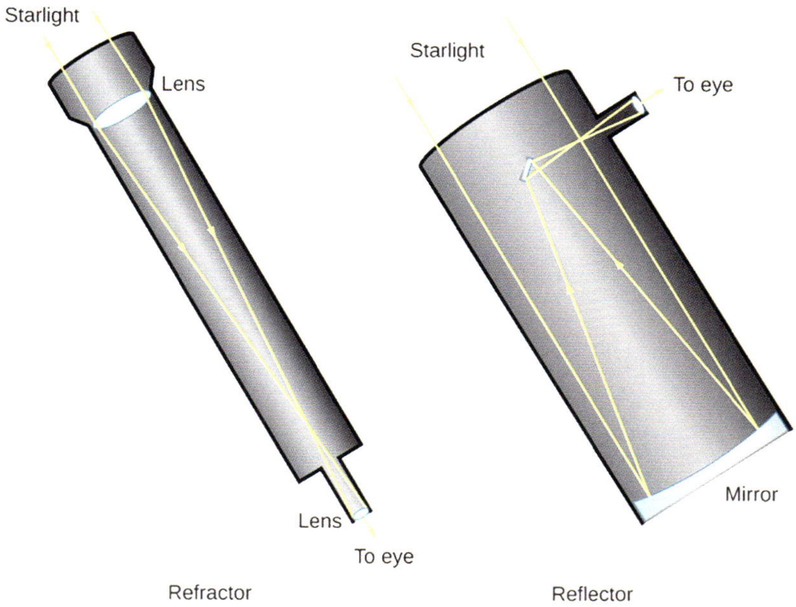

Did You Know?

The James Webb Space Telescope is a huge reflecting telescope. It's located one million miles away from Earth and has a giant mirror that measures almost 22 feet wide.

James Webb Telescope

Chapter 4

The Famous Scientist

In 1671, an organization of influential scientists called the Royal Society of London asked Newton to show them his new reflecting telescope. It was a huge success! After seeing Newton's brilliant invention, the Society invited him to become a member in 1672. He was later named President of the Royal Society in 1703, a role that he held until his death.

Coat of Arms of the Royal Society of London

During Newton's time, the common name for scientists was "natural philosophers." The word "scientist" wasn't even coined until

1834! Although philosophy is much different from science today, it actually comes from Greek words meaning "lover of wisdom." So, someone who gained wisdom about the natural world—like Newton—was called a natural philosopher.

Let There Be Light

In 1672, Newton published his notes on light and color in the Royal Society's journal. By describing how sunlight bends when it passes through a prism, Newton explained that light could be separated into all the colors of the rainbow. He invented the first color wheel, an illustration that includes the seven basic colors: red, orange, yellow,

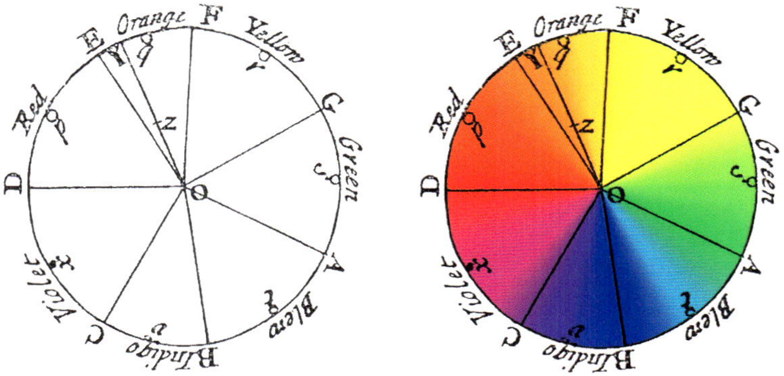

Newton's wheel of color (left) as illustrated in Opticks

green, blue, indigo, and violet. Newton also believed that light was made up of tiny particles known as corpuscles, a theory he would introduce in his famous book on colors and light, *Opticks*, a few decades later in 1704.

In a book published in 1690, Dutch physicist Christiaan Huygens had proposed

"Did blind chance know that there was light and what was its refraction and fit the eyes of all creatures after the most curious manner to make use of it?"

— Isaac Newton

that light is a wave, not a particle. The wave theory was largely accepted and developed by scientists until the early twentieth century. That's when Albert Einstein theorized that light is both a wave and a particle. Newton and Huygens helped light the way for future research.

Isolation and Alchemy

British scientist Robert Hooke

Shortly after Newton joined the Royal Society, the scientist Robert Hooke criticized his findings. Hooke was also a member of the Royal Society, but he had different theories about light and color. He argued that Newton's discoveries weren't reliable and even accused him of plagiarism. These harsh critiques and others deeply affected Newton. Following his mother's death in 1679, he spent the next

The Alchemist *by Mattheus van Helmont*

several years isolated from society.

During this period, Newton began studying alchemy, an ancient practice that formed the basis for modern chemistry. For centuries, alchemists sought to transform common materials into valuable substances, like gold or silver. While the principles of this concept are false, many alchemists learned important things about the elements that scientists research today.

Comets vaporize as they get closer to the sun. This causes them to lose material, which flies into space as a beautiful tail(s).

Chapter 5
A Fresh Start

Though he didn't interact with other scientists much, Newton was dedicated to his own work. After the Great Comet of 1680 passed by Earth, he became interested in astronomy again. Newton used the comet to build upon Johannes Kepler's laws of planetary motion.

Many years before, René Descartes had developed a vortex theory to explain what makes planets orbit. He hypothesized that space itself had currents and moved like water going down a drain. Descartes thought this concept explained orbits, but Newton believed it was the force of gravity that made them possible. With his theory of gravitation, Newton began developing his own mathematical formula to explain why planets move as they do.

"The motions which the planets now have could not spring from any natural cause alone but were [impressed] by an intelligent Agent."

— Isaac Newton

Edmond Halley

In 1684, the astronomer Edmond Halley visited Newton at Cambridge. The young Halley—who later inspired the name of a famous comet—wanted to know if Newton could prove the elliptical orbit of planets. In the months that followed, Newton finalized his calculations until he was able to provide Halley with an answer. He recorded his work in the manuscript *De Motu Corporum in Gyrum*, or *On the Motion of Revolving Bodies*.

Edmond Halley

Principia Mathematica

Halley was so amazed by Newton's findings that he believed the world needed to see them. He encouraged Newton to publish his discoveries. So, Newton combined

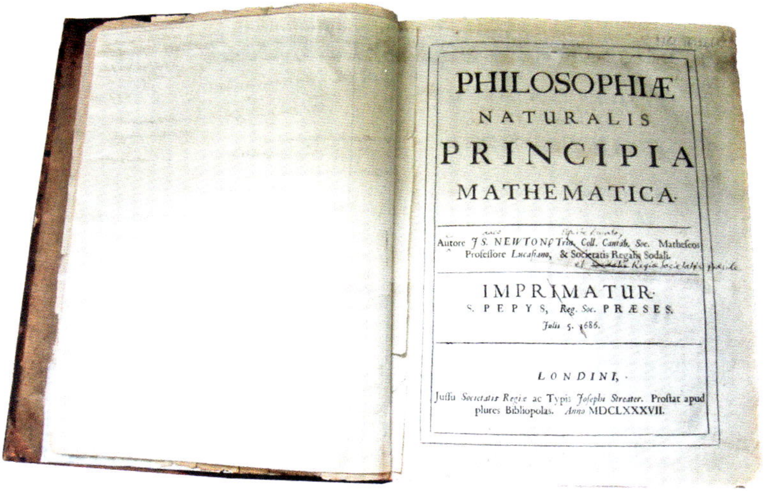

PHILOSOPHIÆ
NATURALIS
PRINCIPIA
MATHEMATICA.

Autore J S. NEWTON, Trin. Coll. Cantab. Soc. Matheseos Professore Lucasiano, & Societatis Regalis Sodali.

IMPRIMATUR.
S. PEPYS, Reg. Soc. PRÆSES.
Julii 5. 1686.

LONDINI,

Jussu Societatis Regiæ ac Typis Josephi Streater. Prostat apud plures Bibliopolas. Anno MDCLXXXVII.

Title page of Principia, *1687*

his years of study on gravity and motion into a single book—*Philosophiæ Naturalis Principia Mathematica*. Published in July 1687, *Principia* is about Newton's theories of physics and astronomy. It contains his three laws of motion, which explain how forces like gravity act upon objects. Even today, *Principia* is considered one of the greatest scientific books ever written.

Did You Know?

The astronomer Edmond Halley helped raise the money for *Principia* to be published. He provided a lot of the funds himself!

SIR ISAAC NEWTON.

Newton's Laws of Motion

The First Law: Inertia

Newton's first law says, "An object in motion will remain in motion, and an object at rest will remain at rest, unless acted upon by an outside force." So, objects keep doing what they're doing unless something outside of them changes that. This concept, known as inertia, describes how movable or unmovable an object is.

Think of a soccer ball, for example. When a soccer player kicks the ball, it starts moving. As the ball moves across the grass, it slows down. That's because the grass puts a force called friction on the ball, causing it to eventually stop.

The Second Law: Acceleration

Newton's second law says, "Net force is equal to mass times acceleration." This law helps us measure how forces affect an object's movement. It's a math formula with three parts:

1) the sum of the forces acting on an object (weight, friction, etc.)
2) an object's mass (inertia)
3) an object's acceleration (change in speed or direction)

So, what does this law look like in action? Imagine you're holding a bowling ball. If you drop the bowling ball from a high place, the ball will move faster as it falls. The force of gravity pulls downward on the mass of the ball, which makes it accelerate.

Did You Know?

The international unit of measure for force is called a newton (N). It's named after Isaac Newton and his work to develop the second law of motion.

The Third Law: Action and Reaction

Newton's third law says, "For every action, there's an equal but opposite reaction." In other words, if you push on an object, the object pushes back on you with an equal force but in the opposite direction. This law is important because it describes how objects interact.

If a baseball player hits a ball with a bat, he can feel the moment the bat hits the ball because the ball exerts a force on the bat. The ball pushes back!

Chapter 6
Other Jobs to Do

In the years after *Principia* was published, Newton didn't focus as much on science. He became involved in politics and represented the University of Cambridge in Parliament. Known for his strong opinions about the English monarchy, Newton was elected to two terms in 1689 and 1701.

1720 engraving of Sir Isaac Newton

Around 1693, Newton suffered from a nervous breakdown. That's when someone becomes so overwhelmed that it's hard to live normally. People aren't quite sure what caused it, but some think it was the result of his chemistry work. In his research, Newton was often exposed to dangerous metals like mercury—a poisonous liquid that can damage the nervous system. After this breakdown, Newton stopped his experiments and turned his attention to other things.

Master of the Mint

Newton served as Warden and then Master of the Royal Mint for over 30 years, starting in 1696. He took his role very seriously and investigated the cases of many illegal counterfeiters—people who make fake money that looks real. Counterfeiters are often able to buy whatever they want with cash they didn't earn.

Queen Anne silver crown coin (one fourth of a pound) minted in 1706

Newton also supervised the recoinage of old silver coins and the production of new coins. He wanted to make sure that each coin was as accurate in weight and fineness of metal as possible. That way, the coins would be more uniform and harder for counterfeiters to replicate.

The Books of Daniel and Revelation

Although he spent a lot of time studying math and science, Newton wrote more notes about the Bible than anything else. He spent much of his life studying God's Word and the works of early Christian leaders.

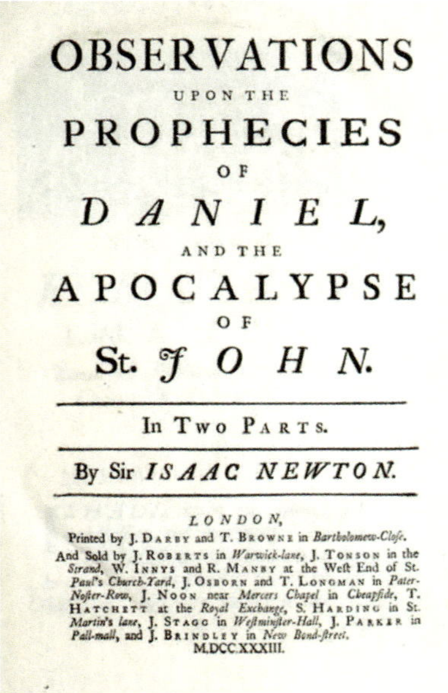

OBSERVATIONS
UPON THE
PROPHECIES
OF
DANIEL,
AND THE
APOCALYPSE
OF
St. *JOHN*.

In Two Parts.

By Sir *ISAAC NEWTON*.

LONDON,
Printed by J. Darby and T. Browne in *Bartholomew-Close*.
And Sold by J. Roberts in *Warwick-lane*, J. Tonson in the *Strand*, W. Innys and R. Manby at the West End of St. *Paul's Church-Yard*, J. Osborn and T. Longman in *Pater-Noster-Row*, J. Noon near *Mercers Chapel* in *Cheapside*, T. Hatchett at the *Royal Exchange*, S. Harding in St. *Martin's lane*, J. Stagg in *Westminster-Hall*, J. Parker in *Pall-mall*, and J. Brindley in *New Bond-street*.
M.DCC.XXXIII.

Newton's book on Daniel and Revelation was published in 1733.

Newton's interest in biblical studies grew as he slowly quit his scientific research. He recorded his thoughts on many topics, but he especially loved reading the books of Daniel in the Old Testament and Revelation in the New Testament. Both of them contain prophecy—Scripture that predicts the future. Newton studied these books often, but he kept his writings about them hidden. They were only published after his death.

> "[God] gave this and the prophecies of the Old Testament, not to gratify men's curiosities by enabling them to foreknow things, but that after they were fulfilled they might be interpreted by the event, and His own providence, not the interpreters, be then manifested thereby to the world."
>
> — Isaac Newton

Knighthood

During the Middle Ages (AD 500–1400), knights were soldiers who fought on horseback. When warfare methods changed, knighthood became a title that honored nonmilitary accomplishments, too. In Newton's time, it was granted to people for service to their country in various fields.

Queen Anne knighted Newton in 1705, so he became known as Sir Isaac Newton. Knighthood is still given by a king or queen today as an award for high achievement.

Sir Isaac Newton

Chapter 7
Impact on the World

When he was an old man, Newton lived for several years with his niece, Catherine Barton, who took care of him and looked after his household. In 1727, Newton died in his sleep at the age of 84. He was the first to be buried in Scientists' Corner in Westminster Abbey—a famous church in London. Other notable scientists, like James Clerk Maxwell and Michael Faraday, are recognized by memorials there.

Newton's tomb in Westminster Abbey

"Diligent, sagacious, and faithful, in his expositions of nature, antiquity, and the holy Scriptures…[Newton] expressed the simplicity of the Gospel in his manners."
— Inscription on Isaac Newton's tomb

Westminster Abbey

The Legacy

1726 portrait of Newton by Enoch Seeman

The legacy of Sir Isaac Newton is still important for us today. His discoveries in physics and calculus provided us with a better understanding of how things work. He learned a lot about the properties of light and the origin of colors. Newton also established a more accurate model of the universe with the telescope he invented. His ideas have changed the way that people view the world around them.

Many believe that Newton had a bigger impact on modern science than anyone else in history. Yet behind his incredible research was something even greater—a steadfast faith in his Creator. Above all, Newton was a scientist who loved God first and His creation, second.

Mysteries of the Universe

Math and science are gifts from our Creator. And He's given us the amazing opportunity to discover them for ourselves! Romans 1:20 says, "For since the creation of the world His invisible attributes are clearly seen, being understood by the things that are made, even His eternal power and Godhead, so that they are without excuse."

Newton was convinced that scientific investigation points to the existence of God and leads to a greater knowledge of Him. What other great mysteries are in God's incredible universe? Maybe you'll discover one of them someday!

"To explain all nature is too difficult a task for any one man or even for any one age. 'Tis much better to do a little with certainty, and leave the rest for others that come after you, than to explain all things by conjecture without making sure of any thing."

— Isaac Newton

A Giant for Others to Stand On

Newton was inspired by the legendary "giants" who came before him. His discoveries influenced future generations of scientists, too. One of them was a Scottish physicist named James Clerk Maxwell. In 1865, Maxwell proved that light, electricity, and magnetism are connected. He discovered that light is made up of interconnected electric and magnetic waves. Maxwell's findings expanded Newton's laws nearly 200 years after they were published.

James Clerk Maxwell

Another scientist who built upon Newton's groundwork was Albert Einstein.

He discovered that gravity isn't a simple force—it's a powerful curve in space. Einstein's study of gravity led him to develop his famous theory of relativity, which changed the world of physics. The principles that Newton developed paved the way for Maxwell, Einstein, and others to make these incredible discoveries and more.

Albert Einstein

$$E=mc^2$$

Albert Einstein's famous formula showing the relationship between energy (E), mass (m), and the speed of light (c)

Timeline of Isaac Newton's Life

1643 – Born in Woolsthorpe, England

1654 – Began school at Free Grammar School in Grantham, England

1661 – Enrolled at Trinity College, University of Cambridge

1665 – Earned bachelor's degree

1665/6 – Discovered gravity and developed calculus

1667 – Became a fellow of Trinity College

1668 – Produced the first reflecting telescope

1669 – Named Lucasian Professor of Mathematics

1672 – Joined the Royal Society of London

1687 – Published *Principia Mathematica*

1689 – Elected to Parliament on behalf of Cambridge

1696/9 – Named Warden then Master of the Royal Mint

1703 – Appointed President of the Royal Society

1705 – Knighted by Queen Anne

1727 – Died and buried in London

Glossary

Astronomy A field of science that involves the study of outer space.

Force A push or pull that acts upon an object.

Optics The study of light, including its properties and behavior.

Physics The study of matter and its movement across time and space.

Plagiarism When someone takes another person's work and claims it as his/her own.

Prism A 3-D shape, usually made of glass, with irregular sides.

Recoinage The making of old coins into new ones.

Relativity The theory that objects can look different based on where you are in the universe.

Scholarship Money that helps pay for someone to attend school.

Successor A person who takes someone else's place after he/she leaves.

Sundial A device that tells the time of day based on the sun's location and shadow.

Image Credits

(t: top; b: bottom; l: left; r: right)

Acabashi via Wikimedia Commons: 12

Yifan Bai via Wikimedia Commons: 14

Bede735 via Wikimedia Commons: 48

BigstockPhoto: 11, 18t, 20, 22, 23, 30, 31r (color portion), 34, 39-41, 43, 46

DeFacto via Wikimedia Commons: 9

Andrew Dunn via Wikimedia Commons: 37

J. Hannan-Briggs via Wikimedia Commons: 10

Andrew Gray via Wikimedia Commons: 21

MostEpic via Wikimedia Commons: 29

NASA: 19, 28

OpenStax via Wikimedia Commons: 27

Javier Otero via Wikimedia Commons: 47

Plucas58 via Wikimedia Commons: 18b

Public domain: 4, 6, 15, 25, 26, 31l, 32-33, 38, 42, 44, 49, 52-54

Vijayakumarblathur via Wikimedia Commons: 51

Wellcome Images via Wikimedia Commons: 16, 36